ANIMAL
TRACKS
— of —
ILLINOIS

Tamara Eder

with contributions from Ian Sheldon

© 2001 by Lone Pine Publishing
First printed in 2001 10 9 8 7 6 5 4 3 2 1
Printed in Canada

All rights reserved. No part of this work covered by the copyright hereon may be
reproduced or used in any form or by any means—graphic, electronic or mechani-
cal—without the prior written permission of the publisher, except for reviewers, who
may quote brief passages. Any request for photocopying, recording, taping or storage
in an information retrieval system of any part of this book shall be directed in writing
to the publisher.

THE PUBLISHER: LONE PINE PUBLISHING

1901 Raymond Avenue SW, Suite C	10145-81 Avenue
Renton, WA 98055	Edmonton, AB T6E 1W9
USA	Canada

Website: http://www.lonepinepublishing.com

National Library of Canada Cataloguing in Publication Data

Eder, Tamara, (date)
  Animal tracks of Illinois

  Includes bibliographical references and index.
  ISBN 1-55105-301-2

  1. Animal tracks—Illinois—Identification. I. Sheldon, Ian, (date) II. Title.
QL768.E34 2001   591.47'9   C2001-910538-X

Editorial Director: Nancy Foulds
Editor: Volker Bodegom
Proofreader: Lee Craig
Production Manager: Jody Reekie
Design, Layout & Production: Volker Bodegom, Monica Triska
Cover Design: Robert Weidemann
Technical Contributor: Mark Elbroch
Cartography: Volker Bodegom
Animal Illustrations: by Gary Ross except for those by Kindrie Grove
   (p. 87), Ted Nordhagen (p. 113), Ewa Pluciennik (p. 109) and Ian
   Sheldon (p. 123)
Track Illustrations: Ian Sheldon
Cover Illustration: Raccoon by Gary Ross
Scanning: Elite Lithographers Ltd.

We acknowledge the financial support of the Government of Canada
through the Book Publishing Industry Development Program (BPIDP)
for our publishing activities.

PC: P4

CONTENTS

How to Use This Book

Most importantly, take this book into the field with you! Relying on your memory is not an adequate way to identify tracks. Track identification has to be done in the field, or with detailed sketches and notes that you can take home. Much of the process of identification involves circumstantial evidence, so you will have much more success when standing beside the track.

This book is laid out in an easy-to-use format. Beginning on p. 126, there is a quick reference appendix to the tracks of all the animals illustrated in the book. This appendix is a fast way to familiarize yourself with certain tracks and the content of the book, and it guides you to the more informative descriptions of each animal and its tracks.

Each description is illustrated with the appropriate footprints and the track patterns that it usually leaves. Although these illustrations are not exhaustive, they do show the tracks or groups of prints that you will most likely see. You will find a list of dimensions for the tracks, giving the general range, but there will always be extremes, just as there are with people who have unusually small or large feet. Under the category 'Size' (of animal), the 'greater-than' sign (>) is used when the size difference between the sexes is pronounced.

If you think that you may have identified a track, check the 'Similar Species' section. This section is designed to help you confirm your conclusions by pointing out other animals that leave similar tracks and showing you ways to distinguish among them.

As you read this book, you will notice an abundance of words such as 'often,' 'mostly' and 'usually.' Unfortunately, tracking will never be an exact science; we cannot expect animals to conform to our expectations, so be prepared for the unpredictable.

Tips on Tracking

As you flip through this guide, you will notice clear, well-formed prints. Do not be deceived! It is a rare track that will ever show so clearly. For a good, clear print, the perfect conditions are slightly wet, shallow snow that is not melting, or slightly soft mud that is not actually wet. These conditions can be rare—most often you will be dealing with incomplete or faint prints where you cannot even really be sure of the number of toes.

Should you find yourself looking at a clear print, then the job of identification is much easier. There are a number of key features to look for: measure the length and width of the print, count the number of toes, check for claw marks and note how far away they are from the body of the print, and look for a heel mark. Keep in mind more subtle features, such as the spacing between the

Plains Pocket Gopher

toes, whether or not they are parallel, and whether fur on the sole of the foot has made the print less clear.

When you are faced with the challenge of identifying an unclear print—or even if you think that you have made a successful identification from one print alone—look beyond the single footprint and search out others. Do not rely on the dimensions of one print alone, but collect measurements from several prints to get an average impression. Even the prints within one trail can show a lot of variation.

Try to determine which is the fore print and which is the hind, and remember that many animals are built very differently from humans, having larger forefeet than

Woodchuck

9

hind feet. Sometimes the prints will overlap, or they can be directly on top of one another in a direct register. For some animals, the fore and hind prints are pretty much the same.

Check out the pattern that the tracks make together in the trail and follow the trail for as many paces as is necessary for you to become familiar with the pattern. Patterns are very important and can be the distinguishing feature between different animals with otherwise similar tracks.

Follow the trail for some distance, because it may give you some vital clues. For example, the trail may lead you to a tree, indicating that the animal is a climber—or it may lead down into a burrow. This part of tracking can be the most rewarding, because you are following the life of the animal as it hunts, runs, walks, jumps, feeds or tries to escape a predator.

Take into consideration the habitat. Sometimes habitat alone will allow you to distinguish very similar tracks— one species might be found on the riverbank, whereas another might be encountered just in dense forest.

Think about your geographical location, too, because some animals have a limited range. This consideration can rule out some species and help you with your identification.

Remember that every animal will at some point leave a print or trail that looks just like the print or trail of a completely different animal!

Finally, keep in mind that if you track quietly, you might catch up with the maker of the prints.

Terms & Measurements

Some of the terms used in tracking can be rather confusing, and they often depend on personal interpretation. For example, what comes to your mind if you see the word 'hopping'? Perhaps you see a person hopping about on one leg—or perhaps you see a rabbit hopping through the countryside. Clearly, one person's perception of motion can be very different from another's.

Mallard

 11

Running: Like galloping, but applied generally to animals moving at high speed. Also used for two-legged animals.

Trotting: Faster than walking, slower than running. The diagonally opposite limbs move simultaneously; that is, the right forefoot with the left hind, then the left forefoot with the right hind. This gait is the natural one for canids (dog family), short-tailed shrews and voles.

hind print

fore print

Canids may use ***side-trotting,*** a fast trotting in which the hind end of the animal shifts to one side. The resulting track pattern has paired tracks, with all the fore prints on one side and all the hind prints on the other.

Walking: A slow gait in which each foot moves independently of the others, resulting in an alternating track pattern. This gait is common for felines (cat family) and deer, as well as wide-bodied animals, such as bears and porcupines. The term is also used for two-legged animals.

Other Tracking Terms:

Dewclaws: Two small, toe-like structures set above and behind the main foot of most hoofed animals.

Direct Register: The hind foot falls directly on the fore print.

double register   *direct register*

Double Register: The hind foot registers and overlaps the fore print only slightly or falls beside it, so that both prints can be seen at least in part.

Dragline: A line left in snow or mud by a foot or the tail dragging over the surface.

dragline

Gallop Group: A track pattern of four prints made at a gallop, usually with the hind feet registering in front of the forefeet (see '***galloping***' for illustration).

Height: Taken at the animal's shoulder.

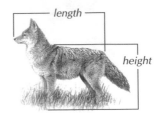

length

height

Length: The animal's body length from head to rump, not including the tail, unless otherwise indicated.

Metacarpal Pad: A small pad near the palm pad or between the palm pad and heel on the forefeet of bears and mustelids (members of the weasel family).

Print (also called '**track**'): Fore and hind prints are treated individually. Print dimensions given are 'length' (including claws—maximum values may represent occasional heel register for some animals) and 'width.' A group of prints made by each of the animal's feet makes up a track pattern.

Register: To leave a mark—said about a foot, claw or other part of an animal's body.

Retractable: Describes claws that can be pulled in to keep them sharp, as with felids (the cat family); these claws do not register in the prints. Foxes have semi-retractable claws.

Sitzmark: The mark left on the ground by an animal falling or jumping from a tree.

Straddle: The total width of the trail, all prints considered.

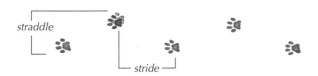

Stride: For consistency among different animals, the stride is taken as the distance from the center of one print (or print group) to the center of the next one. Some books may use the term 'pace.'

Track: Same as '**print**.'

Track Pattern: The pattern left after each foot registers once; a set of prints, such as a gallop group.

Trail: A series of track patterns; think of it as the path of the animal.

Beaver

MAMMALS

River Otter

White-tailed Deer

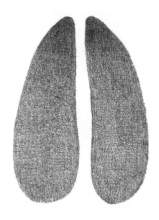

Fore and Hind Prints
Length: 2–3.5 in (5–9 cm)
Width: 1.6–2.5 in (4–6.5 cm)

Straddle
5–10 in (13–25 cm)

Stride
Walking: 10–20 in (25–50 cm)
Galloping: 6–15 ft (1.8–4.5 m)

Size (buck>doe)
Height: 3–3.5 ft (90–110 cm)
Length: to 6.3 ft (1.9 m)

Weight
120–350 lb (55–160 kg)

walking   *gallop group*

WHITE-TAILED DEER
Odocoileus virginianus

The keen hearing of this deer guarantees that it knows about you before you know about it. Frequently, all that we see is its conspicuous white tail in the distance as it gallops away, earning this deer the nickname 'Flagtail.' This adaptable deer may be found throughout the state, in small groups at the edges of forests and in brushlands. The White-tailed Deer can be common around ranches and residential areas.

This deer's prints are heart-shaped and pointed. Its alternating walking track pattern shows the hind prints direct registered or double registered on the fore prints. When a deer gallops on a soft surface, such as snow, the dewclaws register. This flighty deer gallops in the usual style, leaving hind prints ahead of fore prints, with the toes splayed wide for steadier, safer footing.

Similar Species: No other wild cloven-hoofed animals inhabit this region, but in farming areas the prints of domesticated Elk (*Cervus elaphus*) or the Domestic Pig (*Sus scrofa*) could be mistaken for deer prints.

Black Bear

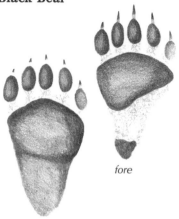

fore

hind

Fore Print
Length: 4–6.3 in (10–16 cm)
Width: 3.8–5.5 in (9.5–14 cm)

Hind Print
Length: 6–7 in (15–18 cm)
Width: 3.5–5.5 in (9–14 cm)

Straddle
9–15 in (23–38 cm)

Stride
Walking: 17–23 in (43–58 cm)

Size (male>female)
Height: 3–3.5 ft (90–110 cm)
Length: 5–6 ft (1.5–1.8 m)

Weight
200–600 lb (90–270 kg)

walking

BLACK BEAR
Ursus americanus

The Black Bear has a scattered range in this part of the country, and although there is no longer a breeding population in Illinois, infrequent reports of wandering bears still occur. Finding fresh bear tracks can be a thrill, but take care—the bear may be just ahead. Never underestimate the potential power of a surprised bear!

Black Bear prints somewhat resemble small human prints, but wider and with claw marks. The small inner toe rarely registers. The forefoot's small heel pad often registers, and the hind print shows a big heel. The bear's slow walk results in a slightly pigeon-toed double register with the hind print on the fore print. More frequently, at a faster pace, the hind foot oversteps the forefoot. When a bear runs, the two hind feet register in front of the forefeet in an extended cluster. Along well-worn bear paths, look for 'digs' (patches of dug-up earth) and 'bear trees' whose scratched bark shows that this bear climbs.

Similar Species: No other wild animal is likely to leave similar tracks in Illinois.

Coyote

fore

hind

Fore Print
(hind print is slightly smaller)
Length: 2.4–3.2 in (6–8 cm)
Width: 1.6–2.4 in (4–6 cm)

Straddle
4–7 in (10–18 cm)

Stride
Walking: 8–16 in (20–40 cm)
Trotting: 17–23 in (43–58 cm)
Galloping/Leaping:
    2.5–10 ft (0.8 m–3 m)

Size (female is slightly smaller)
Height: 23–26 in (58–65 cm)
Length: 32–40 in (80–100 cm)

Weight
20–50 lb (9–23 kg)

walking or
trotting

gallop
group

COYOTE
(Brush Wolf, Prairie Wolf)
Canis latrans

This widespread, adaptable canine prefers open grasslands or woodlands. On its own, with a mate or in a family pack, it hunts rodents and larger prey. If you find a coyote den—usually a wide-mouthed tunnel leading into a nesting chamber—do not bother the family or the female will have to move her pups to a safer location.

The hind print is slightly smaller than the oval fore print, and its less triangular heel pad rarely registers clearly. The claws of the two outer toes usually do not register. The Coyote typically walks or trots in an alternating pattern; the walk has a wider straddle, and the trotting trail is often very straight. When the Coyote gallops, its hind feet fall in front of its forefeet; the faster it goes, the straighter the gallop group. The Coyote's tail, which hangs down, leaves a dragline in deep snow.

Similar Species: A Domestic Dog's (p. 26) less oval prints splay more, and its trail will be erratic. Foot hairs blur Red Fox (p. 30) prints (usually smaller). Gray Fox (p. 32) prints are much smaller.

Gray Fox

fore

hind

Fore Print (hind print is slightly smaller)
Length: 1.3–2.1 in (3.3–5.3 cm)
Width: 1.1–1.5 in (2.8–3.8 cm)

Straddle
2–4 in (5–10 cm)

Stride
Walking/Trotting: 7–12 in (18–30 cm)

Size
Height: 14 in (35 cm)
Length: 21–30 in (53–75 cm)

Weight
7–15 lb (3.2–7 kg)

walking

GRAY FOX
Urocyon cinereoargenteus

This small, shy fox is widespread but quite rare; look for its tracks in woodlands and chaparral country. The Gray Fox is the only fox that climbs trees, which it does either for safety or to forage.

The forefoot registers better than the smaller hind foot, and the hind foot's long, semi-retractable claws do not always register. The heel pads are often unclear—they sometimes show up just as small, round dots. When it walks, this fox leaves a neat alternating track pattern. When it trots, its prints fall in pairs, with the fore print set diagonally behind the hind print. The Gray Fox's gallop group is like the Coyote's (p. 28).

Similar Species: The Red Fox's (p. 30) fore heel pad has a bar across it; in general, Red Fox prints are larger and less clear (because of thick fur), and they have a longer stride and narrower straddle. Coyote tracks are much larger. Domestic Dog (p. 26) prints are much larger. Feline (pp. 34–37) prints lack claw marks, and they have larger, less symmetrical heel pads.

Bobcat

fore

hind

Fore Print
(hind print is slightly smaller)
Length: 1.8–2.5 in (4.5–6.5 cm)
Width: 1.8–2.5 in (4.5–6.5 cm)

Straddle
4–7 in (10–18 cm)

Stride
Walking: 8–16 in (20–40 cm)
Running: 4–8 ft (1.2–2.4 m)

Size (female is slightly smaller)
Height: 20–22 in (50–55 cm)
Length: 25–30 in (65–75 cm)

Weight
15–35 lb (7–16 kg)

walking

ambling to loping

BOBCAT
(Wildcat)
Lynx rufus

The handsome
Bobcat, a stealthy
and usually noc-
turnal hunter,
is seldom seen.
Once numerous in Illinois, it is now on the endangered
species list. You might encounter a solitary wanderer anywhere from wilderness to residential areas.

A walking Bobcat's hind feet usually register directly
on its larger fore prints. As the Bobcat picks up speed,
its trail becomes an ambling pattern of paired prints, the
hind leading the fore. At even greater speeds, it leaves
four print groups in a lope pattern. Especially the fore
prints show asymmetry. The front part of the heel pad
has two lobes and the rear part has three. In deep snow
the Bobcat's feet leave draglines. Half-buried scat along
the Bobcat's meandering trail marks its territory.

Similar Species: A large Domestic Cat (p. 36) will make
similar prints, but it will have a shorter stride and a narrower straddle; a Domestic Cat will not wander far from
home, especially in winter. Fox (pp. 30–33), Domestic
Dog (p. 26) and Coyote (p. 28) prints are narrower than
long and show claw marks, and the fronts of their foot-
pads are once-lobed. Weasel (pp. 46–49) or Mink (p. 44)
prints that show four toes may also look similar.

Raccoon

fore

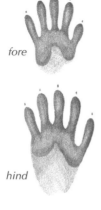

hind

Fore Print
Length: 2–3 in (5–7.5 cm)
Width: 1.8–2.5 in (4.5–6.5 cm)

Hind Print
Length: 2.4–3.8 in (6–9.5 cm)
Width: 2–2.5 in (5–6.5 cm)

Straddle
3.3–6 in (8.5–15 cm)

Stride
Walking: 8–18 in (20–45 cm)
Bounding: 15–25 in (38–65 cm)

Size (female is slightly smaller)
Length: 24–37 in (60–95 cm)

Weight
11–35 lb (5–16 kg)

walking

bounding group

RACCOON
Procyon lotor

The inquisitive Raccoon is common throughout Illinois. Adored by some people for its distinctive face mask, it is disliked for its boundless curiosity—often demonstrated with residential garbage cans. Look for tracks near water at low elevations. The Raccoon likes to rest in trees. It usually dens up in cold weather.

The Raccoon's unusual print, showing five well-formed toes, looks like a human handprint; its small claws make dots. Its highly dexterous forefeet rarely leave heel prints, but its hind prints, which are generally much clearer, do show heels. The Raccoon's peculiar walking track pattern shows the left fore print next to the right hind print (or just in front) and vice versa. If a Raccoon is out in deep snow, it may use a direct-registering walk. The Raccoon occasionally bounds, leaving clusters with the two hind prints in front of the fore prints.

Similar Species: Unclear Opossum (p. 40) prints may look similar, but the Opossum drags its tail. In mud or wet snow, River Otter (p. 42) tracks may look similar. Woodchuck (p. 60) prints may also look similar.

Opossum

fore

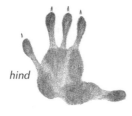

hind

Fore Print
Length: 2–2.3 in (5–5.8 cm)
Width: 2–2.3 in (5–5.8 cm)
Hind Print
Length: 2.5–3 in (6.5–7.5 cm)
Width: 2–3 in (5–7.5 cm)
Straddle
4–5 in (10–13 cm)
Stride
5–11 in (13–28 cm)
Size
Length: 2–2.5 ft (60–75 cm)
Weight
9–13 lb (4–6 kg)

walking   *fast walking*

OPOSSUM
Didelphis virginiana

This slow-moving, nocturnal marsupial is found throughout Illinois. It occupies many habitats and is quite tolerant of residential areas, but it prefers open woodland or brushland around waterbodies. Look for Opossum tracks in mud near the water and in snow during the warmer months of winter (the Opossum dens up during freezing weather) or near roadkill (which it likes to eat), though many Opossums suffer the same fate as the carrion that they dine on.

The Opossum has two walking habits: the common alternating pattern, with the hind prints registering on the fore prints, and a Raccoon-like (p. 38) paired-print pattern, with each hind print next to the opposing fore print. The very distinctive, long, inward-pointing thumb of the hind foot does not make a claw mark. The Opossum is an excellent climber. In snow, the long, naked tail's dragline may be bloodstained; not well adapted to cold, this thinly haired animal frequently suffers frostbite.

Similar Species: Prints in which the distinctive thumbs do not show may be mistaken for a Raccoon's.

Mink

fore

hind

Fore and Hind Prints
Length: 1.3–2 in (3.3–5 cm)
Width: 1.3–1.8 in (3.3–4.5 cm)

Straddle
2.1–3.5 in (5.3–9 cm)

Stride
Walking/2×2 loping: 8–36 in (20–90 cm)

Size (male>female)
Length with tail: 19–28 in (48–70 cm)

Weight
1.5–3.5 lb (0.7–1.6 kg)

2×2 loping

MINK
Mustela vison

The lustrous Mink, found scattered throughout Illinois, prefers watery habitats surrounded by brush or forest. At home as much on land as in water, this nocturnal hunter can be exciting to track. Like the River Otter (p. 42), the Mink slides in mud and snow, carving out a trough up to 6 inches (15 cm) wide.

The Mink's fore print shows five (perhaps four) toes, with five loosely connected palm pads in an arc, but the hind print shows only four palm pads. The metacarpal pad of the forefoot rarely registers, but the furred heel of the hind foot may register, lengthening the hind print. The Mink prefers the typical mustelid 2×2 loping gait, which leaves consistently spaced, slightly angled double prints. Its diverse track patterns also include alternating walking, loping with three- and four-print groups (like the River Otter) and bounding (like a rabbit, p. 54).

Similar Species: The Long-tailed Weasel (p. 46) makes similar but generally smaller tracks. The River Otter's trail shows distinct tail draglines. Bobcat (p. 34) prints may resemble four-toed Mink prints, but they will not show claw marks or lobed palm pads, and they will not be in a bounding pattern.

45

Long-tailed Weasel

Fore and Hind Prints
Length: 1.1–1.8 in (2.8–4.5 cm)
Width: 0.8–1 in (2–2.5 cm)

Straddle
1.8–2.8 in (4.5–7 cm)

Stride
Bounding: 9.5–43 in (24–110 cm)

Size (male>female)
Length with tail: 12–22 in (30–55 cm)

Weight
3–9.5 oz (85–270 g)

2×2 loping

LONG-TAILED WEASEL
Mustela frenata

The widely distributed Long-tailed Weasel is an active year-round hunter that has an avid appetite for rodents. Following this nimble creature's tracks can reveal much about its activities. Some weasel trails may lead you up a tree. Weasels sometimes take to water. Tracks are most evident in winter, when weasels frequently burrow into the snow or pursue rodents into their holes. In deep snow, look for holes where the weasel has suddenly plunged in, perhaps in pursuit of prey.

The usual weasel gait is a 2×2 lope, leaving a trail of paired prints. The Long-tailed Weasel's typical 2×2 lope shows an irregular stride—sometimes short and sometimes long—with no consistent behavior. Like the Mink (p. 44), this weasel may bound like a rabbit (p. 54).

Similar Species: The Least Weasel (p. 48) makes similar but smaller tracks. Mink also make similar tracks, and small Mink tracks may be indistinguishable from large Long-tailed Weasel tracks.

Least Weasel

Fore and Hind Prints
Length: 0.5–0.8 in (1.3–2 cm)
Width: 0.4–0.5 in (1–1.3 cm)

Straddle
0.8–1.5 in (2–3.8 cm)

Stride
2×2 loping: 5–20 in (13–50 cm)

Size (male>female)
Length with tail: 6.5–9 in (17–23 cm)

Weight
1.3–1.8 oz (37–51 g)

2×2 loping

LEAST WEASEL
Mustela nivalis

The Least Weasel is Illinois's smallest weasel, and it makes the least-clear tracks—you might find them in northern Illinois around wetlands and in open woodlands and fields. If you find a good Least Weasel trail, following it may turn into an obstacle course, because this nimble weasel runs over and under logs, up and down trees and in and out of holes.

Because of this weasel's light weight and small, hairy feet, the pad detail is often obscured, especially in dry dusty soil or flaky snow. Even with clear tracks, the inner toe rarely registers.

Similar Species: Close attention to stride and straddle is useful in identification, because a large Least Weasel track can look very similar to a small or juvenile Long-tailed Weasel (p. 46) track. Another good way to help differentiate these two weasels is by weight—the heavier Long-tailed Weasel makes deeper and clearer tracks.

Badger

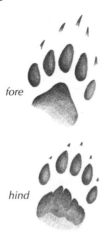

fore

hind

Fore Print
(hind print is slightly shorter)
Length: 2.5–3 in (6.5–7.5 cm)
Width: 2.3–2.8 in (5.8–7 cm)

Straddle
4–7 in (10–18 cm)

Stride
Walking: 6–12 in (15–30 cm)

Size
Length: 21–36 in (53–90 cm)

Weight
13–25 lb (6–11 kg)

walking

BADGER
Taxidea taxus

The Badger, with its squat shape and unmistakable face, is most often seen in the open grasslands, but the Badger also ventures into higher country. Thick shoulders and forelegs, coupled with long claws, make it a powerful digger. Look for the Badger's tracks in northern Illinois, particularly in spring and fall snow—unlike most other mustelids, it likes to den up in a hole for the coldest months of winter.

When a Badger walks, the alternating track pattern shows a double register, with the hind print sometimes falling just behind (or sometimes slightly in front of) the fore print. All five toes on each foot register. A Badger's long claws are evident in the pigeon-toed track that it leaves as it waddles along; the forefoot claws are longer than the hind-foot ones. In deep snow, the plowing action of the Badger's wide, low body often wipes out track detail.

Similar Species: Very large Striped Skunk (p. 52) prints may appear similar, but skunk tracks rarely show a consistent pattern, and their toes do not turn in as much.

Eastern Cottontail

fore

hind

Fore Print
Length: 1–1.5 in (2.5–3.8 cm)
Width: 0.8–1.3 in (2–3.3 cm)
Hind Print
Length: 3–3.5 in (7.5–9 cm)
Width: 1–1.5 in (2.5–3.8 cm)
Straddle
4–5 in (10–13 cm)
Stride
Hopping: 0.6–3 ft (18–90 cm)
Size
Length: 12–17 in (30–43 cm)
Weight
1.3–3 lb (0.6–1.4 kg)

hopping

EASTERN COTTONTAIL
Sylvilagus floridanus

This abundant rabbit is widespread throughout Illinois. Preferring brushy areas in grasslands and cultivated areas, it might be found in dense vegetation, hiding from predators such as the Bobcat (p. 34) and the Coyote (p. 28). Largely nocturnal, the Eastern Cottontail might be seen at dawn or dusk and on darker days.

As with other rabbits and hares, this rabbit's most common track pattern is a triangular grouping of four prints, with the larger hind prints (which can appear pointed) falling in front of the fore prints (which may overlap). The hairiness of the toes will hide any pad detail. If you follow this rabbit's trail, you could be startled if it flies out from its 'form,' a depression in the ground in which it rests.

Similar Species: The Swamp Rabbit (*S. aquaticus*), which makes similar but larger tracks, lives in the southern part of the state, usually in the vicinity of water. Tree squirrel (pp. 66–73) tracks show a similar pattern, but the fore prints are more consistently side by side.

Beaver

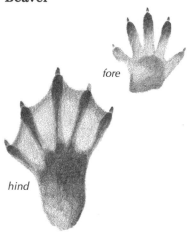

fore

hind

Fore Print
Length: 2.5–4 in (6.5–10 cm)
Width: 2–3.5 in (5–9 cm)
Hind Print
Length: 5–7 in (13–18 cm)
Width: 3.3–5.3 in (8.5–13 cm)
Straddle
6–11 in (15–28 cm)
Stride
Walking: 3–6.5 in (7.5–17 cm)
Size
Length with tail: 3–4 ft (90–120 cm)
Weight
28–75 lb (13–34 kg)

walking

BEAVER
Castor canadensis

Few animals leave as many signs of their presence as the Beaver, which is capable of changing the local landscape. North America's largest rodent, it is a common sight around water. Look for its conspicuous dams and lodges and for the stumps of felled trees. Check trunks gnawed clean of bark for marks of the Beaver's huge incisors. Scent mounds marked with castoreum, a strong-smelling yellowish fluid that Beavers produce, also indicate recent activity.

Check the large hind prints for signs of webbing and broad toenails. The nail of the fourth toe usually does not register, and it is rare for all five toes on each foot to do so. Irregular foot placement in the alternating walking gait may produce a direct register or a double register. The Beaver's thick, scaly tail may mar its tracks, as can the branches that it drags about for construction and food. Repeated path use results in well-worn trails.

Similar Species: The Beaver's many signs and large hind prints minimize confusion. Muskrat (p. 58) prints are much smaller.

Muskrat

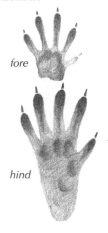

fore

hind

Fore Print
Length: 1.1–1.5 in (2.8–3.8 cm)
Width: 1.1–1.5 in (2.8–3.8 cm)
Hind Print
Length: 1.6–3.2 in (4–8 cm)
Width: 1.5–2.1 in (3.8–5.3 cm)
Straddle
3–5 in (7.5–13 cm)
Stride
Walking: 3–5 in (7.5–13 cm)
Running: to 1 ft (30 cm)
Size
Length with tail: 16–25 in (40–65 cm)
Weight
2–4 lb (0.9–1.8 kg)

walking

MUSKRAT
Ondatra zibethicus

Like the Beaver (p. 56), this rodent is found throughout Illinois, wherever there is water. Beavers are very tolerant of Muskrats and even allow them to live in parts of their lodges. Active all year, the Muskrat leaves plenty of signs. It digs extensive networks of burrows, often undermining riverbanks, so do not be surprised if you suddenly fall into a hidden hole! Also look for small lodges in the water and beds of vegetation on which the Muskrat rests, suns and feeds in summer.

The small fifth (innermost) toe of the forefoot rarely registers. Stiff hairs that aid in swimming may create a 'shelf' around the five well-formed toes of the hind print. The common alternating walking pattern shows print pairs that alternate from side to side; the hind print is just behind the fore print or slightly overlaps it. In snow, a Muskrat's feet drag, and its tail leaves a sweeping dragline.

Similar Species: Few animals share this water-loving rodent's habits. The Beaver makes larger tracks, and it leaves many other signs as well.

Woodchuck

fore

hind

Fore and Hind Prints
Length: 1.8–2.8 in (4.5–7 cm)
Width: 1–2 in (2.5–5 cm)

Straddle
3.3–6 in (8.5–15 cm)

Stride
Walking: 2–6 in (5–15 cm)
Bounding: 6–14 in (15–35 cm)

Size (male>female)
Length with tail:
    20–25 in (50–65 cm)

Weight
5.5–12 lb (2.5–5.5 kg)

walking   *bounding*

WOODCHUCK
(Whistle Pig,
Groundhog,
Marmot)
Marmota monax

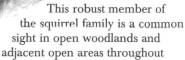

This robust member of
the squirrel family is a common
sight in open woodlands and
adjacent open areas throughout
Illinois. Always on the watch for predators, but not too
troubled by humans, the Woodchuck never wanders far
from its burrow. This marmot hibernates during winter
but emerges in early spring; look for tracks in late spring
snowfalls and in mud around the burrow entrances.

A Woodchuck's fore print shows four toes, three palm
pads and two heel pads (not always evident). The hind
print shows five toes, four palm pads and two poorly
registering heel pads. The Woodchuck usually leaves an
alternating walking pattern, with the hind print registered
on the fore print. When a Woodchuck runs from danger,
it makes groups of four prints, hind ahead of fore.

Similar Species: Squirrel (pp. 62, 66–71) prints are very
similar, but they are smaller. A small Raccoon's (p. 38)
bounding track pattern will be similar, but it will show
five-toed fore prints.

Thirteen-lined Ground Squirrel

fore

hind

Fore Print
Length: 1–1.3 in (2.5–3.3 cm)
Width: 0.5–1 in (1.3–2.5 cm)

Hind Print
Length: 1.1–1.5 in (2.8–3.8 cm)
Width: 0.8–1.3 in (2–3.3 cm)

Straddle
2.3–3.5 in (5.8–9 cm)

Stride
Bounding: 7–20 in (18–50 cm)

Size
Length with tail: 7–12 in (18–30 cm)

Weight
4–10 oz (110–280 g)

bounding

THIRTEEN-LINED GROUND SQUIRREL
(Striped Gopher)
Spermophilus tridecemlineatus

With its many stripes and dots, this adorable ground squirrel of the shortgrass areas is an attractive sight. Its range includes all of Illinois.

This animal's tracks may be evident near its many burrow entrances in mud or in late season snowfalls. If there are no fresh tracks around the burrow, the animal may be dormant. The small fifth toe of the forefoot rarely registers in the fore print, and the two heel pads sometimes show. The larger hind print shows five toes. Both forefeet and hind feet have long claws that frequently register. Usually seen scurrying around, ground squirrels leave a typical squirrel track pattern, with the hind prints registering ahead of the fore prints, which are usually placed diagonally.

Similar Species: Franklin's Ground Squirrel (*S. franklinii*) is larger. Chipmunk (p. 64) tracks are smaller. Tree squirrel (pp. 66–73) bounding groups have more of a square shape. Pocket gopher (p. 78) tracks tend to be smaller and more elongated.

Eastern Chipmunk

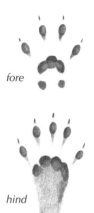

fore

hind

Fore Print
Length: 0.8–1 in (2–2.5 cm)
Width: 0.4–0.8 in (1–2 cm)

Hind Print
Length: 0.7–1.3 in (1.8–3.3 cm)
Width: 0.5–0.9 in (1.3–2.3 cm)

Straddle
2–3.2 in (5–8 cm)

Stride
Running: 7–15 in (18–38 cm)

Size
Length with tail: 7–10 in (18–25 cm)

Weight
2.5–5 oz (70–140 g)

bounding

EASTERN CHIPMUNK
Tamias striatus

Look for this delightful character throughout the state. The Eastern Chipmunk is found in a variety of habitats, from the dense forest floor to open areas near buildings. You are more likely to see or hear this character, which is highly active during summer, than to notice its tracks. This large chipmunk is happiest on the ground, but it will gladly climb sturdy oak trees to harvest juicy, ripe acorns. It enters a deep sleep in winter, waking up from time to time to eat.

Chipmunks are so light that their tracks rarely show fine details. The forefeet each have four toes and the hind feet have five. Chipmunks run on their toes, so the two heel pads of the forefeet seldom register; the hind feet have no heel pads. Their erratic track patterns, like those of many of their cousins, show the hind feet registering in front of the forefeet. A chipmunk trail often leads to extensive burrows.

Similar Species: No other chipmunks live in this state. Tree squirrels (pp. 66–73) usually have larger prints and a wider straddle, and they are more likely to make midwinter tracks. Mouse (pp. 82–85) tracks are smaller.

Eastern Gray Squirrel

fore

hind

Fore Print
Length: 1–1.8 in (2.5–4.5 cm)
Width: 1 in (2.5 cm)

Hind Print
Length: 2.3–3 in (5.8–7.5 cm)
Width: 1.1–1.5 in (2.8–3.8 cm)

Straddle
3.8–6 in (9.5–15 cm)

Stride
Bounding: 0.7–3 ft (22–90 cm)

Size
Length with tail: 17–20 in (43–50 cm)

Weight
14–25 oz (400–710 g)

bounding

EASTERN GRAY SQUIRREL
Sciurus carolinensis

This large, familiar squirrel can be a common sight in deciduous and mixed forests throughout Illinois, even in urban areas. Active all year, the Eastern Gray Squirrel can leave a wealth of evidence, especially in winter as it scurries about digging up nuts that it buried during the previous fall.

The Eastern Gray Squirrel leaves a typical squirrel track pattern when it runs or bounds. The hind prints fall slightly in front of the fore prints. A clear fore print shows four toes with sharp claws, four fused palm pads and two heel pads. The hind print shows five toes and four palm pads; if the full heel-length registers, it also shows two small heel pads.

Similar Species: Fox Squirrel (p. 68) prints are as big or larger. Red Squirrel (p. 70) prints are smaller. Chipmunks (p. 64) and flying squirrels (p. 72) make smaller tracks in a similar pattern, but they have narrower straddles. Rabbits (p. 54) make longer track patterns, and their forefeet rarely register side by side when they run.

Fox Squirrel

fore

hind

Fore Print
Length: 1–1.9 in (2.5–4.5 cm)
Width: 1–1.7 in (2.5–4.3 cm)

Hind Print
Length: 2–3.3 in (5–7.5 cm)
Width: 1.5–1.9 in (3.8–4.8 cm)

Straddle
4–6 in (10–15 cm)

Stride
Bounding: 0.7–3 ft (22–90 cm)

Size
Length with tail: 18–28 in (45–70 cm)

Weight
1–2.4 lb (0.5–1.1 kg)

bounding

FOX SQUIRREL
Sciurus niger

This squirrel is much like the Eastern Gray Squirrel (p. 66), but it is larger and has a yellowish underside. It can be a common sight in deciduous forests with plenty of nut trees and in open areas or woodlands throughout Illinois. Piles of nutshells at tree bases indicate its favorite feeding sites. Active all year, the Fox Squirrel spends a lot of time foraging on the ground, often collecting nuts that it buried singly during the previous fall.

A clear fore print shows four toes with claws evident, four fused palm pads and two heel pads. The hind print shows five toes, four palm pads and sometimes a heel. When it runs or bounds, the Fox Squirrel makes a typical squirrel track, the hind prints slightly in front of the fore prints, with the prints in each pair roughly side by side.

Similar Species: Eastern Gray Squirrel prints are generally slightly smaller. Chipmunk (p. 64), Red Squirrel (p. 70) and flying squirrel (p. 72) tracks fall in a similar pattern, but they are smaller, and the straddles are narrower. Rabbits (p. 54) make longer print groups; their fore prints rarely register side by side when they run.

Red Squirrel

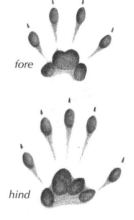

fore

hind

Fore Print
Length: 0.8–1.5 in (2–3.8 cm)
Width: 0.5–1 in (1.3–2.5 cm)

Hind Print
Length: 1.5–2.3 in (3.8–5.8 cm)
Width: 0.8–1.3 in (2–3.3 cm)

Straddle
3–4.5 in (7.5–11 cm)

Stride
Bounding: 8–30 in (20–75 cm)

Size
Length with tail:
    9–15 in (23–38 cm)

Weight
2–9 oz (57–260 g)

bounding

bounding (deep snow)

RED SQUIRREL
(Pine Squirrel, Chickaree)
Tamiasciurus hudsonicus

When you
enter a Red
Squirrel's terri-
tory, the inhabitant
greets you with a
loud, chattering call.
Another obvious sign of this
forest dweller, which is found
in northern Illinois, is its large
middens—piles of cone scales and cores
left beneath trees—that indicate favorite feeding sites.

Active all year in its small territory, a Red Squirrel
will leave an abundance of trails that lead from tree to
tree or down a burrow. This energetic animal mostly
bounds, leaving groups of four prints, the hind prints in
front of the fore prints, which tend to be side by side (but
not always). Four toes show on each fore print, and five
show on each hind print. The heels often do not register
when squirrels move quickly. In deeper snow the prints
merge to form pairs of diamond-shaped tracks.

Similar Species: The larger Fox Squirrel (p. 68) and
Eastern Gray Squirrel (p. 66) both make similar but
larger tracks. Eastern Chipmunk (p. 64) and Southern
Flying Squirrel (p. 72) tracks fall in a similar pattern,
but they are smaller and have narrower straddles.

Southern Flying Squirrel

fore

hind

Fore Print
Length: 0.3–0.5 in (0.8–1.3 cm)
Width: 0.4 in (1 cm)

Hind Print
Length: 0.9–1.3 in (2.3–3.3 cm)
Width: 0.5 in (1.4 cm)

Straddle
2–2.5 in (5–6.5 cm)

Stride
Bounding: 7–22 in (18–55 cm)

Size
Length with tail: 8–10 in (20–25 cm)

Weight
1.5–3.2 oz (43–90 g)

bounding

SOUTHERN FLYING SQUIRREL
Glaucomys volans

This soft-furred brown acrobat is capable of long-distance gliding using the membranes that connect its forelegs and hindlegs. Primarily nocturnal, the Southern Flying Squirrel is found in coniferous and mixed forests throughout the state. Up to 50 of these squirrels may huddle together in a nest for warmth in winter, but they do not truly hibernate.

When it is on the ground, this flying squirrel normally bounds. Though its hind prints may register in front of its fore prints, New England-based tracking expert Mark Elbroch reports that its more common pattern (with a narrower straddle than that of other squirrels) shows the fore prints in front. If it glides down to the ground, this squirrel can leave a distinctive four-print 'sitzmark' in loose material or snow, but it often climbs down a tree trunk instead.

Similar Species: Other tree squirrels (pp. 66–71) usually make larger prints, and they rarely leave sitzmarks, but with tracks in loose dirt or snow it can be impossible to identify the squirrel species. Chipmunk (p. 64) tracks are smaller, and they have a narrower straddle.

Eastern Woodrat

fore

hind

Fore Print
Length: 0.6–0.8 in (1.5–2 cm)
Width: 0.4–0.5 in (1–1.3 cm)

Hind Print
Length: 1–1.5 in (2.5–3.8 cm)
Width: 0.6–0.8 in (1.5–2 cm)

Straddle
2.3–2.8 in (5.8–7 cm)

Stride
Walking: 1.8–3 in (4.5–7.5 cm)
Bounding: 5–8 in (13–20 cm)

Size
Length with tail:
    12–17 in (30–43 cm)

Weight
8–16 oz (230–450 g)

walking   *bounding*

EASTERN WOODRAT

Neotoma floridana

   This woodrat, considered
endangered in Illinois, may be found in
rocky areas in the southernmost part of the
state. If you are lucky enough to encounter a trail
of this woodrat, it might lead you to a distinctive mass
of a nest, most often in a crevice, shrub or burrow.
Although it favors rocky areas, it feeds on foliage, seeds,
ferns and fungi and therefore requires nearby vegetation.

   Four toes show on the fore print and five on the hind
The short claws rarely register. A woodrat often walks in
an alternating fashion, with the hind print direct registering on the fore print. This woodrat frequently bounds as
well, leaving a pattern of four prints, with the larger hind
print in front of the diagonally placed fore prints. The
stride tends to be short relative to the size of the prints.

Similar Species: Indistinct squirrel (pp. 62, 66–73)
prints are similar but usually larger. The Norway Rat
(p. 76) has similar prints, but it is usually found close
to human activity. Woodchuck (p. 60) prints are similar
but much larger.

Norway Rat

fore

hind

Fore Print
Length: 0.7–0.8 in (1.8–2 cm)
Width: 0.5–0.7 in (1.3–1.8 cm)

Hind Print
Length: 1–1.3 in (2.5–3.3 cm)
Width: 0.8–1 in (2–2.5 cm)

Straddle
2–3 in (5–7.5 cm)

Stride
Walking: 1.5–3.5 in (3.8–9 cm)
Bounding: 9–20 in (23–50 cm)

Size
Length with tail: 13–19 in (33–48 cm)

Weight
7–18 oz (200–510 g)

walking

NORWAY RAT
(Brown Rat)
Rattus norvegicus

Active both day and
night, this despised rat is widespread almost anywhere that humans
have decided to build their homes. Not entirely
dependent on people, it may live in the wild as well.

The fore print shows four toes, and the hind print
shows five. When it bounds, this colonial rat leaves four-
print groups, with the hind prints in front of the diagonally placed fore prints. Sometimes one of the hind feet
direct registers on a fore print, creating a three-print
group. This rat more commonly leaves an alternating
walking pattern with the larger hind prints close to or
overlapping the fore prints; the hind heel does not show.
The tail often leaves a dragline in loose material. Rats
live in groups, so you may find many trails together,
often leading to their 5-inch (2-cm) wide burrows.

Similar Species: The Eastern Woodrat (p. 74), which is
very rare in Illinois, leaves similar tracks, but it seldom
associates with human activity. Mouse (pp. 82–85) prints
are much smaller. Red Squirrel (p. 70) tracks show distinctive squirrel traits. The Eastern Chipmunk (p. 64)
usually bounds, and its tracks are often smaller.

Plains Pocket Gopher

fore

hind

Fore Print
Length: 1 in (2.5 cm)
Width: 0.6 in (1.5 cm)
Hind Print
Length: 0.8–1 in (2–2.5 cm)
Width: 0.5 in (1.3 cm)
Straddle
1.5–2 in (3.8–5 cm)
Stride
Walking: 1.3–2 in (3.3–5 cm)
Size (male> female)
Length with tail: 6–9 in (15–23 cm)
Weight
2.8–5 oz (80–140 g)

walking

PLAINS POCKET GOPHER

Geomys bursarius

This seldom-seen rodent of central Illinois spends most of its time in burrows, venturing out only to move mud around and to find a mate. Because of its need to dig, the Plains Pocket Gopher prefers soft, moist soils, and it especially enjoys pastureland.

By far the best sign of pocket gopher activity is the muddy mounds that it creates when it burrows. As the gopher digs, it pushes the excess soil up out of the hole, creating a mound. Then, when it has finished digging, it pushes up a bit more soil to plug the entrance or 'cap' the burrow. Search around the mounds to find tracks. Each foot has five toes. Though the forefeet have long, well-developed claws for digging, the prints rarely show this much detail. Pocket gophers usually walk, leaving an alternating track pattern in which the hind prints fall on or slightly behind the fore prints.

Similar Species: Pocket gopher tracks are associated with their distinctive burrows. The Eastern Mole (p. 88) also leaves piles of pushed-up soil (usually smaller), but it does not make distinct plugs. Moles also leave ridges. Ground squirrels (p. 62) tend to be larger, and they leave rounder tracks with four-toed fore prints.

Woodland Vole

fore

hind

Fore Print
Length: 0.5 in (1.3 cm)
Width: 0.5 in (1.3 cm)

Hind Print
Length: 0.6 in (1.5 cm)
Width: 0.5–0.8 in (1.3–2 cm)

Straddle
1.3–2 in (3.3–5 cm)

Stride
Walking/Trotting: 0.8 in (2 cm)
Bounding: 2–6 in (5–15 cm)

Size
Length with tail:
    4–5.5 in (10–14 cm)

Weight
0.8–1.3 oz (23–37 g)

walking

*bounding
(in snow)*

WOODLAND VOLE
(Pine Vole)
Microtus pinetorum

Distinguishing a vole track from those of the wealth of other small mammals in the state can be challenging. If you do spot a vole track, throughout Illinois the most likely candidate is the Woodland Vole. It lives in a variety of habitats.

When clear (which is seldom), vole fore prints show four toes, and hind prints show five. A vole's walk and trot both leave a paired alternating track pattern with a hind print occasionally direct registered on a fore print. Voles usually opt for a faster bounding in which the hind prints register on the fore prints to form print pairs. This vole lopes quickly across open areas, creating a three-print track pattern. Voles stay under the snow in winter; when it melts, look for distinctive piles of cut grass from their ground nests. The bark at the bases of shrubs may show tiny teeth marks left by gnawing. In summer, well-used vole paths appear as little runways in the grass.

Similar Species: Other voles include the Prairie Vole (*M. ochrogaster*), which is common in dry, grassy areas, and the Meadow Vole (*M. pennsylvanicus*), which is found in northern Illinois. Mouse (pp. 82–85) bounding tracks show four-print groups.

White-footed Mouse

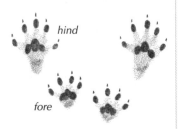

hind

fore

bounding group

Fore Print
Length: 0.3–0.4 in (0.8–1 cm)
Width: 0.3–0.4 in (0.8–1 cm)

Hind Print
Length: 0.3–0.5 in (0.8–1.3 cm)
Width: 0.3–0.4 in (0.8–1 cm)

Straddle
1.4–1.8 in (3.5–4.5 cm)

Stride
Bounding: 3–12 in (7.5–30 cm)

Size
Length with tail:
    6–12 in (15–30 cm)

Weight
0.5–1.3oz (14–35 g)

bounding

*bounding
(in snow)*

WHITE-FOOTED MOUSE

Peromyscus leucopus

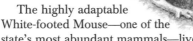

The highly adaptable White-footed Mouse—one of the state's most abundant mammals—lives anywhere from woodlands to cultivated fields. It is seldom seen, because it is nocturnal. This mouse may enter buildings in winter, where it will stay active. In colder areas it may hibernate.

In perfect, soft mud, the fore prints each show four toes, three palm pads and two heel pads, and the hind prints show five toes and three palm pads; the hind heel pads rarely register. Bounding tracks, most noticeable in snow, show the hind prints falling in front of the fore prints. In flaky snow the prints may merge to look like larger pairs of prints; tail drag will be evident. This mouse's trail may lead up a tree or down into a burrow.

Similar Species: The Deer Mouse (*P. maniculatus*), which is also common, makes identical prints. The House Mouse (*Mus musculus*), with similar tracks, associates more with humans. Jumping mouse (p. 84) prints show long, thin toes. Voles (p. 80) tend to trot, and they have a much shorter bounding track pattern. Chipmunks (p. 64) have a wider straddle. Shrews (p. 86) have a narrower straddle.

Meadow Jumping Mouse

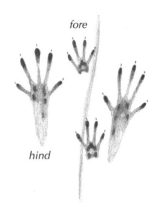

fore

hind

Fore Print
Length: 0.3–0.5 in (0.8–1.3 cm)
Width: 0.3–0.5 in (0.8–1.3 cm)

Hind Print
Length: 0.5–1.3 in (1.3–3.3 cm)
Width: 0.5–0.7 in (1.3–1.8 cm)

Straddle
1.8–1.9 in (4.5–4.8 cm)

Stride
Bounding: 7–18 in (18–45 cm)
In alarm: 3–6 ft (90–180 cm)

Size
Length with tail: 7–9 in (18–23 cm)

Weight
0.6–1.3 oz (17–35 g)

bounding

MEADOW JUMPING MOUSE

Zapus hudsonius

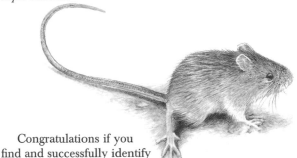

Congratulations if you find and successfully identify the tracks of the Meadow Jumping Mouse! Though it can be locally abundant, its preference for grassy meadows and its long, deep winter hibernation (about six months!) make locating tracks very difficult.

Jumping mouse tracks are distinctive if you do find them. The two smaller fore prints are between the long hind prints; the long heels do not always register, and some prints show just the three long middle toes. The toes on the forefeet may splay so much that the side toes point backward. When they bound, jumping mice make short leaps. The tail may leave a dragline in soft mud or unseasonable snow. Clusters of cut grass stems about 5 inches (13 cm) long found lying in meadows are a more abundant sign of this rodent.

Similar Species: Though this mouse's prints are normally distinctive, heel-less hind prints may be mistaken for a vole's (p. 80) or a small bird's (pp. 110–113) or an amphibian's (pp. 114–119).

Masked Shrew

hind

fore

*bounding
group*

Fore Print
Length: 0.2 in (0.5 cm)
Width: 0.2 in (0.5 cm)
Hind Print
Length: 0.6 in (1.5 cm)
Width: 0.3 in (0.8 cm)
Straddle
0.8–1.3 in (2–3.3 cm)
Stride
Bounding: 1.2–2.3 in (3–5.8 cm)
Size
Length with tail: 3–4 in (7.5–10 cm)
Weight
0.1–0.2 oz (3–6 g)

bounding

MASKED SHREW
Sorex cinereus

Though several species of tiny, frenetic shrews live in Illinois, the abundant Masked Shrew is a likely candidate if you find tracks. This shrew prefers moist fields, bogs, marshes or woodlands, but it can also be found in higher and drier grasslands. Its rapid activity makes it difficult to observe closely.

In its energetic and unending quest for food, a shrew usually leaves a four print bounding pattern, but it may slow to an alternating walking pattern. The individual prints in a group are often indistinct but, in mud or shallow, wet snow, you can even count the five toes on each print. In deeper snow a shrew's tail often leaves a dragline. If a shrew tunnels under snow, it may leave a snow ridge on the surface. A shrew's trail may disappear down a burrow.

Similar Species: The Southeastern Shrew (*S. longirostris*) and the Least Shrew (*Cryptotis parva*) have indistinguishable tracks. Mouse (pp. 82–85) fore prints show four toes.

Eastern Mole

a molehill of the Eastern Mole

some molehills and ridges of the Eastern Mole

Size
Length: 4.5–6.5 in (11–17 cm)
Tail length: 1–1.5 in (2.5–3.8 cm)
Weight
2.5–5 oz (70–140 g)

EASTERN MOLE
Scalopus aquaticus

This soft-furred resident of the underworld is the only mole that you are likely to encounter in Illinois. It can certainly leave a wealth of evidence to indicates its presence, usually in pastures or open woodlands, especially where the soil is light and moist and easy to burrow in.

Moles, who seldom emerge from their subterranean environment, create an extensive network of burrows through which they forage. These burrows are sometimes marked by ridges on the surface, though most of us are more familiar with the hills that form where the mole gets rid of excess soil from its burrows and for which moles are frequently considered to be pests when they mess up fine lawns. When rains moisten the soil and bring worms to the surface, it can be entertaining to watch the earth twitch and rise up as the mole satisfies its voracious appetite.

Similar Species: No other moles are common in this state. The Plains Pocket Gopher (p. 78) may push up similar piles, but they will generally be larger and have distinct plugs that 'cap' the holes themselves.

BIRDS, AMPHIBIANS & REPTILES

A guide to the animal tracks of Illinois is not complete without some consideration of the birds, amphibians and reptiles found in this state.

Several bird species have been chosen to represent the main types common to this region, but remember that individual bird species are not easily identified by track alone. The shores of lakes and streams are very reliable places to find bird tracks—the mud there can hold a clear print for a long time. The sheer number of tracks made by shorebirds and waterfowl can be astonishing. Bird tracks are also abundant in snow and are clearest in shallow, wet snow. Though some bird species prefer to perch in trees or soar across the sky, it can be entertaining to follow the tracks of birds that spend a lot of time on the ground. They can spin around in circles and lead you in all directions. The trail may suddenly end as the bird takes flight, or it might terminate in a pile of feathers, the bird having fallen victim to a predator.

Many amphibians and turtles depend on moist environments, so look in the soft mud along the shores of lakes and ponds for their distinctive tracks. You may be able to distinguish frog tracks from toad tracks, because these amphibians generally move differently, but it can be very difficult to identify the species. Reptiles thrive and outnumber the amphibians in drier environments, but they seldom leave good tracks, except in occasional mud or perhaps in sand. Snakes leave distinctive body prints.

 91

Mallard

Print
Length: 2–2.5 in (5–6.5 cm)
Straddle
4 in (10 cm)
Stride
to 4 in (10 cm)
Size
23 in (58 cm)

MALLARD
Anas platyrhynchos

male

female

This dabbling duck—the male a familiar sight with its striking green head—is common in open areas near lakes and ponds. Its webbed feet leave prints that can often be seen in abundance along the muddy shores of just about any waterbody, including those in urban parks.

The webbed foot of the Mallard has three long toes that all point forward. Though the toes register well, the webbing between the toes does not always show in the print. The inward-pointing feet give the Mallard a pigeon-toed appearance, which perhaps accounts for its waddling gait, a characteristic for which ducks are known.

Similar Species: Many waterfowl, such as other ducks, as well as the Herring Gull (p. 94), leave similar prints. Exceptionally large prints were likely made by a goose (various species) or a swan (*Cygnus* spp.).

Herring Gull

Print
Length: 3.5 in (9 cm)
Straddle
4–6 in (10–15 cm)
Stride
4.5 in (11 cm)
Size
Length: 23–25 in (58–65 cm)

HERRING GULL
Larus argentatus

The Herring Gull, with its long wings and webbed toes, is a strong long-distance flier as well as an excellent swimmer. Common in a variety of habitats, it can be found concentrated in great numbers along waterbodies.

Gulls leave slightly asymmetrical tracks that show three toes. They have claws that register outside the webbing, and the claw marks are usually attached to the footprint. Most gulls have quite a swagger to their gait, and they leave a trail with the tracks turned strongly inward.

Similar Species: Gull species cannot be reliably identified by track alone, but smaller species have conspicuously smaller tracks. Mallard (p. 92) and other duck tracks are often difficult to distinguish from gull tracks. Geese (various species) make larger prints.

Great Blue Heron

Print
Length: to 6.5 in (17 cm)
Straddle
8 in (20 cm)
Stride
9 in (23 cm)
Size
4.2–4.5 ft (1.3–1.4 m)

GREAT BLUE HERON
Ardea herodias

The refined and graceful image of this large heron symbolizes the precious wetlands in which it patiently hunts for food. Usually still and statuesque as it waits for a meal to swim by, the Great Blue Heron will have cause to walk from time to time, perhaps to find a better hunting location. Look for its large, slender tracks along the banks or mudflats of waterbodies.

Not surprisingly, a bird that lives and hunts with such precision walks in a similar fashion, leaving straight tracks that fall in a nearly straight line. Look for the slender rear toe in the print.

Similar Species: Other herons and egrets—such as the Green Heron (*Butorides virescens*), the Great Egret (*Casmerodius albus*) and the night herons (*Nycticorax* spp.)—have similar prints that vary in size relative to the size of the bird.

Common Snipe

Print
Length: 1.5 in (3.8 cm)
Straddle
to 1.8 in (4.5 cm)
Stride
to 1.3 in (3.3 cm)
Size
11–12 in (28–30 cm)

COMMON SNIPE
Gallinago gallinago

   This short-legged character is a resident of marshes and bogs, where its neat prints can often be seen in mud. Snipes are quite secretive when on the ground, and so you may be surprised if one suddenly flushes out from beneath your feet. If there is a Common Snipe in the air, you may hear an eerie whistle if it dives from the sky.

   The Common Snipe's neat prints show four toes, including a small rear toe that points inward. The bird's short legs and stocky body give it a very short stride.

Similar Species: Many shorebirds, including the Spotted Sandpiper (p. 100), leave similar tracks.

Spotted Sandpiper

Print
Length: 0.8–1.3 in (2–3.3 cm)
Straddle
to 1.5 in (3.8 cm)
Stride
Erratic
Size
7–8 in (18–20 cm)

SPOTTED SANDPIPER
Actitis macularia

The bobbing tail of the Spotted Sandpiper is a common sight on the shores of lakes, rivers and streams, but you will usually find just one of these territorial birds in any given location. Because of its excellent camouflage, likely the first that you will see of this bird will be when it flies away, its fluttering wings close to the surface of the water.

As a sandpiper teeters up and down on the shore, it leaves trails of three-toed prints that show a very small fourth toe facing off to one side at an angle. Sandpiper tracks can have an erratic stride.

Similar Species: All sandpipers and plovers, such as the common Killdeer (*Charadrius vociferus*), leave similar tracks, although there is much diversity in size. The Common Snipe (p. 98) makes similar but larger tracks.

Ruffed Grouse

Print
Length: 2–3 in (5–7.5 cm)
Straddle
0.5–2 in (1.3–5 cm)
Stride
Walking: 3–6 in (7.5–15 cm)
Size
15–19 in (38–48 cm)

RUFFED GROUSE
Bonasa umbellus

This ground-dweller prefers the quiet seclusion of forests, where its excellent camouflage usually affords it good protection. Reintroductions of this native bird, once considered extirpated in Illinois, mean that you have increasing chances of encountering its tracks. Although you might find its tracks in mud, snow much improves your chances of finding them. If you follow a Ruffed Grouse trail quietly, you may be startled when the bird bursts from cover almost beneath your feet.

The three thick front toes leave very clear impressions, but the short rear toe, which is angled off to one side, does not always show up so well. This bird's neat, straight trail appears to reflect its cautious approach to life on the forest floor.

Similar Species: The Greater Prairie Chicken (*Tympanuchus cupido*)—endangered in the state—leaves similar tracks. The Wild Turkey (*Meleagris gallopavo*), also reintroduced, leaves similar but much larger tracks.

Great Horned Owl

Strike
Width: to 3 ft (90 cm)
Size
22 in (55 cm)

GREAT HORNED OWL
Bubo virginianus

Often seen resting quietly in trees by day, this wide-ranging owl prefers to hunt at night. Signs of this bird are more common in winter, when the owl often strikes through the snow with its talons, leaving an untidy hole that may be surrounded by wing and tail-feather imprints. If it registers well, this 'strike' can be quite a sight. The feather imprints are made as the owl struggles to take off with possibly heavy prey. An ungraceful walker, it prefers to fly away from the scene.

You may stumble across a strike and guess that the owl's target could have been a vole (p. 80) scurrying around underneath the snow. Or you may be following the surface trail of an animal to find that it abruptly ends with this strike mark where the animal has been seized.

Similar Species: If the prey left no approaching trail (meaning that it was moving under the snow), the strike mark is likely an owl's, because owls hunt by sound. If there is a trail, the strike mark is probably from a bird of prey that hunts by sight instead.

American Crow

Print
Length: 2.5–3 in (6.5–7.5 cm)
Straddle
1.5–3 in (3.8–7.5 cm)
Stride
Walking: 4 in (10 cm)
Size
16 in (40 cm)

AMERICAN CROW
Corvus brachyrhyncos

The black silhouette of the American Crow can be a common sight in a variety of habitats. The American Crow will frequently come down to the ground and contentedly strut around with a confidence that hints at its intelligence. Its loud *caw* can be heard from quite a distance. Crows can be especially noisy when they are mobbing an owl or a hawk.

The American Crow typically leaves an alternating walking track pattern. Its prints show three sturdy toes pointing forward and one toe pointing backward. When a crow is in need of greater speed, perhaps for take-off, it bounds along, leaving irregular pairs of diagonally placed prints with a longer stride between each pair.

Similar Species: Other corvids, such as the Blue Jay (*Cyanocitta cristata*), also spend a lot of time on the ground and make similar tracks that vary in size depending on the size of the bird.

Northern Flicker

Print
Length: 1.8 in (4.5 cm)
Straddle
1–1.5 in (2.5–3.8 cm)
Stride
Hopping: 1.5–5 in (3.8–13 cm)
Size
5.5–6.5 in (14–17 cm)

NORTHERN FLICKER
Colaptes auratus

male

female

This attractive woodpecker,
which can be seen throughout Illinois,
is common in open woodlands and right into suburban
areas. Unlike most woodpeckers, it spends some of its
time feeding on the ground.

A clear flicker track shows a distinctive arrangement
of two strong toes pointing forward and two pointing to
the rear, with the outer toes slightly longer than the inner
ones. The flicker's toes—along with short, strong legs that
give the bird a short stride—are well suited for grasping
tree trunks and limbs as this agile bird works its way
along in search of insects.

Similar Species: Most other birds have very different
tracks. Other woodpeckers would leave similar tracks,
but very few come down to the ground as much as the
obliging Northern Flicker does.

Northern Cardinal

Print
Length to: 1.5 in (3.8 cm)
Straddle
1–1.5 in (2.5–3.8 cm)
Stride
Hopping: 1.5–5 in (3.8–13 cm)
Size
9 in (23 cm)

NORTHERN CARDINAL
Cardinalis cardinalis

female

male

     The brilliant red plumage and small black mask and chin of the male are a joy to see on this year-round resident. Like the Dark-eyed Junco (p. 110), the Northern Cardinal is a small hopping bird.

Each foot has three forward-pointing toes and one longer toe at the rear. The best prints are left in mud or snow, although in loose, flaky snow the toe detail is lost, and the feet may show some dragging between the hops.

A good place to study these types of prints is near a birdfeeder. Watch the birds scurry around as they pick up fallen seeds, then have a look at the prints that they have left behind.

Similar Species: Juncos, finches (various species) and sparrows (various species) make similar tracks. The size of the toes may indicate what kind of bird you are tracking—larger birds have larger footprints. Not all birds are present all year, so keep in mind the season when tracking.

Frogs

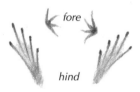

fore

hind

Straddle
to 3 in (7.5 cm)

hopping

FROGS

Bullfrog

The best place to look for frog tracks is along the muddy fringes of waterbodies. A frog generally hops, resulting in its two small forefeet registering in front of its long-toed hind prints. Frog tracks vary greatly in size, depending on species and age.

Of the many frogs in Illinois, the treefrogs are among the smallest. The Spring Peeper's (*Pseudacris crucifer*) 1.5-inch (3.8-cm) length and its preference for thick undergrowth and shrubs near the water make its tracks a rare sight. The larger gray treefrogs (*Hyla chrysoscelis* and *H. versicolor*) spend most of their time in trees, coming down to breed and sing at night. The widespread Green Frog (*Rana clamitans*) favors slow-moving, shallow water and swampy areas. In the southern half of Illinois, look for the larger tracks of the beautiful Southern Leopard Frog (*R. sphenocephala*), which grows to 5 inches (13 cm) long. Unusually large tracks are surely from the robust Bullfrog (*R. catesbeiana*). Growing to 8 inches (20 cm) in length, it is North America's largest frog.

Toads

hind *fore*

Straddle
to 2.5 in (6.5 cm)

walking

TOADS

American Toad

Undoubtedly, the best place to look for toad tracks is, as with frog tracks, along the muddy fringes of waterbodies, but they can occasionally be found in drier areas—for example, as unclear trails in dusty patches of soil. In general, toads walk and frogs (p. 114) hop, but toads are pretty capable hoppers too, especially when being hassled by overly enthusiastic naturalists. Toads leave rather abstract prints as they walk. The heels of the hind feet do not register. On less firm surfaces, the toes often leave draglines.

There are fewer toad species than frog species in Illinois. The toad most likely to be encountered, and the most widespread, is the American Toad (*Bufo americanus*); it lives in many different moist habitats. Fowler's Toad (*B. fowleri*) is found scattered throughout most of Illinois in temporary pools and ditches. The Eastern Spadefoot (*Scaphiopus holbrookii*) can be found in forests, brushy areas or cultivated land in the southernmost parts of the state. Toads in Illinois can be up to 4.5 inches (11 cm) in length.

Lizards

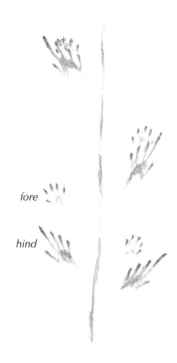

fore

hind

Straddle
to 3 in (7.5 cm)

walking

LIZARDS

Five-lined Skink

Lizards can be difficult to find in Illinois, but there are several species that inhabit the southern part of the state. Their tracks are similar to salamander tracks, but they show longer, more slender toes. These reptiles all move very quickly when the need arises, their feet barely touching the ground as they dart for cover. Consequently, their tracks can be hard to make out clearly.

If you find lizard tracks, a likely candidate is the Five-lined Skink (*Eumeces fasciatus*), a lizard that can grow to 8 inches (20 cm) in length. It favors moist woodlands, but it is sometimes found in moist cultivated areas. In the south you might also encounter the Broadhead Skink (*E. laticeps*). This skink also favors moist woodlands, and if you find its tracks you can probably also find the skink by sorting through the nearest pile of leaf litter or lifting up the nearest piece of wood. The Racerunner (*Cnemidophorus sexlineatus*), which is rare in Illinois, favors dry grasslands and well-drained woodlands; it can reach 11 inches (28 cm) in length.

Turtles

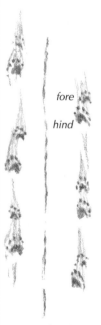

fore

hind

fore

hind

Straddle
4–10 in (10–25 cm)

Snapping Turtle walking

typical turtle walking

122

TURTLES

Painted Turtle

Turtles, those ancient inhabitants of the water world, will happily slip into the murky depths to avoid detection. They do, however, come out from time to time to feed or to bask in the sunshine. Look for their distinctive tracks alongside ponds, rivers and moist areas. Some turtles, such as the huge Common Snapping Turtle (*Chelydra serpentina*), prefer to stay in the water and rarely come out.

With its large shell and short legs, a turtle leaves a track that is wide relative to the length of its stride—its straddle is about half its body length. Although longer-legged turtles can raise their shells off the ground, short-legged species may let them drag, as shown in their tracks. The tail may leave a straight dragline in the mud. On firmer surfaces, look for distinct claw marks. One of the most widespread and prettiest turtles, often seen basking, is the Painted Turtle (*Chrysemys picta*); it can reach almost 10 inches (25 cm) in length. Slightly smaller is the well-named Common Musk Turtle (*Sternotherus odoratus*), which lives in shallow water and produces a foul odor. The Slider (*Trachemys scripta*), which reaches 12 inches (30 cm) in length, prefers slow-moving rivers in the south of the state. The Eastern Box Turtle (*Terrapene carolina*) is common in forested areas in the southern half of the state.

Snakes

SNAKES

Common Garter Snake

Many snake species inhabit Illinois, with much greater diversity in the warmer south. Because snakes are all long and slender, their tracks appear so similar that identification among the species is next to impossible. In fact, because a snake lacks feet and leaves a track that is just a gentle meander, it is very challenging even to establish in which direction a snake was moving.

The harmless Common Garter Snake (*Thamnophis sirtalis*) is the most frequently encountered snake. Found throughout the state, often near wet or moist areas, it can reach 4.3 feet (1.3 m) in length. Also widespread, but to only 16 inches (40 cm) in length, the Redbelly Snake (*Storeria occipitomaculata*) prefers hilly woodlands. The Rough Green Snake (*Opheodrys aestivus*), which can grow to 3.8 feet (1.2 m) long, inhabits grassy meadows and fields along forest edges in the state's southern half. The rattlesnake most likely to be encountered is the Massasauga (*Sistrurus catenatus*), which can grow to 3.3 feet (1 m) long and frequents a variety of habitats from marshlands to dry woodlands. Also common in a variety of habitats is the large and colorful Milk Snake (*Lampropeltis triangulum*), which grows up to 6.5 feet (2 m) long.

TRACK PATTERNS & PRINTS

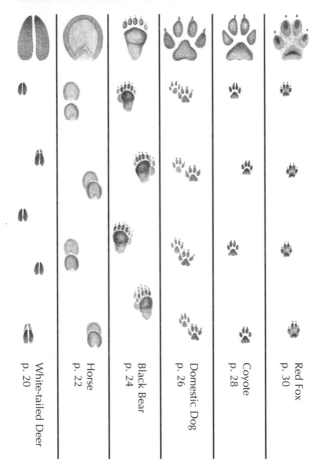

White-tailed Deer
p. 20

Horse
p. 22

Black Bear
p. 24

Domestic Dog
p. 26

Coyote
p. 28

Red Fox
p. 30

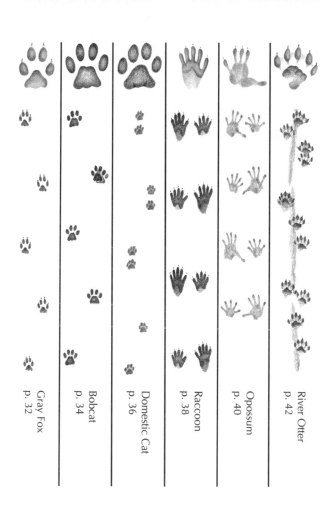

Gray Fox
p. 32

Bobcat
p. 34

Domestic Cat
p. 36

Raccoon
p. 38

Opossum
p. 40

River Otter
p. 42

TRACK PATTERNS & PRINTS

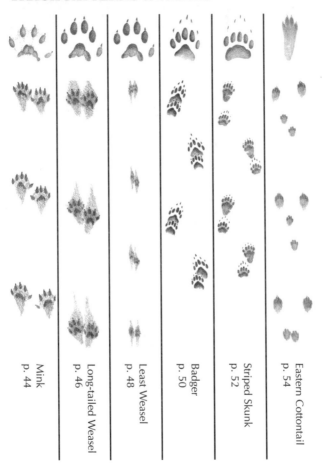

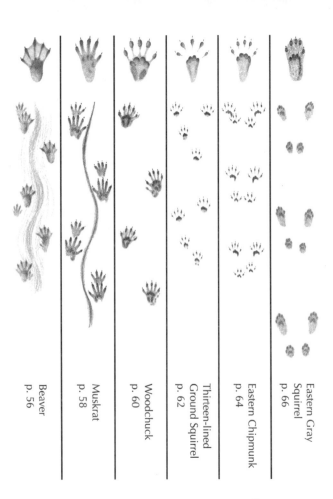

Beaver
p. 56

Muskrat
p. 58

Woodchuck
p. 60

Thirteen-lined
Ground Squirrel
p. 62

Eastern Chipmunk
p. 64

Eastern Gray
Squirrel
p. 66

129

TRACK PATTERNS & PRINTS

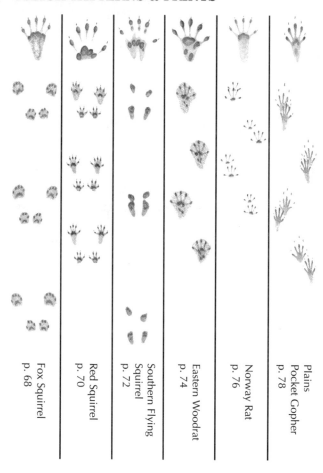

Fox Squirrel
p. 68

Red Squirrel
p. 70

Southern Flying
Squirrel
p. 72

Eastern Woodrat
p. 74

Norway Rat
p. 76

Plains
Pocket Gopher
p. 78

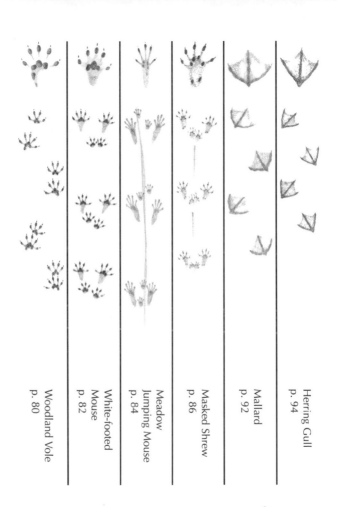

TRACK PATTERNS & PRINTS

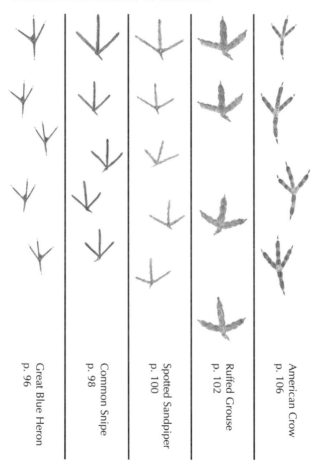

Great Blue Heron
p. 96

Common Snipe
p. 98

Spotted Sandpiper
p. 100

Ruffed Grouse
p. 102

American Crow
p. 106

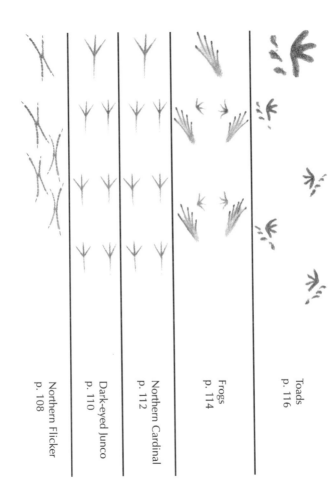

133

TRACK PATTERNS & PRINTS

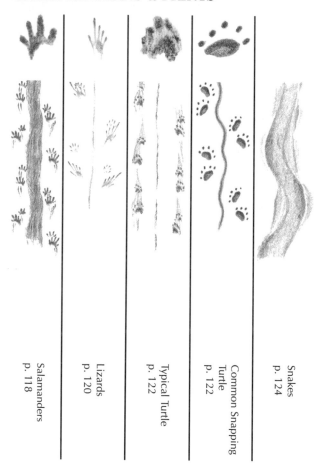

HOOFED PRINTS

White-tailed
Deer

Horse

inch cm

0 — 0

1

2 — 5

HIND PRINTS

White-
footed
Mouse

Masked
Shrew

Woodland
Vole

Plains
Pocket
Gopher

inch cm

0 — 0

1

1 — 2

3

4

2 — 5

Southern
Flying Squirrel

Eastern
Chipmunk

Meadow
Jumping Mouse

135

HIND PRINTS

Norway
Rat

Thirteen-lined
Ground Squirrel

Eastern
Woodrat

Red
Squirrel

Woodchuck

Muskrat

Fox
Squirrel

Eastern Gray
Squirrel

inch cm
0 — 0
1
2 — 5

Opossum

Raccoon

Eastern
Cottontail

HIND PRINTS

inch cm
0 — 0
2
4 — 10

Beaver

Black Bear

FORE PRINTS

Least
Weasel

Long-tailed
Weasel

Mink

Striped
Skunk

River
Otter

Badger

Domestic
Cat

Gray
Fox

Red
Fox

Bobcat

Coyote

Domestic Dog

inch cm
0 — 0

1

2 — 5

137

BIBLIOGRAPHY

Behler, J.L., and F.W. King. 1979. *Field Guide to North American Reptiles and Amphibians.* National Audubon Society. New York: Alfred A. Knopf.

Brown, R., J. Ferguson, M. Lawrence and D. Lees. 1987. *Tracks and Signs of the Birds of Britain and Europe: An Identification Guide.* London: Christopher Helm.

Burt, W.H. 1976. *A Field Guide to the Mammals.* Boston: Houghton Mifflin Company.

Farrand, J., Jr. 1995. *Familiar Animal Tracks of North America.* National Audubon Society Pocket Guide. New York: Alfred A. Knopf.

Forrest, L.R. 1988. *Field Guide to Tracking Animals in Snow.* Harrisburg: Stackpole Books.

Halfpenny, J. 1986. *A Field Guide to Mammal Tracking in North America.* Boulder: Johnson Publishing Company.

Headstrom, R. 1971. *Identifying Animal Tracks.* Toronto: General Publishing Company.

Murie, O.J. 1974. *A Field Guide to Animal Tracks.* The Peterson Field Guide Series. Boston: Houghton Mifflin Company.

Rezendes, P. 1992. *Tracking and the Art of Seeing: How to Read Animal Tracks and Signs.* Vermont: Camden House Publishing.

Stall, C. 1989. *Animal Tracks of the Rocky Mountains.* Seattle: The Mountaineers.

Stokes, D., and L. Stokes. 1986. *A Guide to Animal Tracking and Behaviour.* Toronto: Little, Brown and Company.

Wassink, J.L. 1993. *Mammals of the Central Rockies.* Missoula: Mountain Press Publishing Company.

Whitaker, J.O., Jr. 1996. *National Audubon Society Field Guide to North American Mammals.* New York: Alfred A. Knopf.

INDEX

Page numbers in **boldface** type refer to the primary (illustrated) treatments of animal species and their tracks.

ABOUT THE AUTHORS

Tamara Eder, equipped from the age of six with a canoe, a dip net and a note pad, grew up with a fascination for nature and the diversity of life. She has a degree in environmental conservation sciences and has photographed and written about the biodiversity in Bermuda, the Galapagos Islands, the Amazon Basin, China, Tibet, Vietnam, Thailand and Malaysia.

Ian Sheldon, an accomplished artist, naturalist and educator, has lived in South Africa, Singapore, Britain and Canada. Caught collecting caterpillars at the age of three, he has been exposed to the beauty and diversity of nature ever since. He was educated at Cambridge University and the University of Alberta. When he is not in the tropics working on conservation projects or immersing himself in our beautiful wilderness, he is sharing his love for nature. Ian enjoys communicating this passion through the visual arts and the written word.